# The High-Quality Care

## *Practical Tips for Improving Compliance*

# The Path to High-Quality Care

## *Practical Tips for Improving Compliance*

American Animal Hospital Association
12575 West Bayaud Avenue
Lakewood, Colorado 80228
800/252-2242 or 303/986-2800
*www.aahanet.org*



Printed in the United States of America

**Library of Congress Cataloging-in-Publication Data**

The path to high quality care : practical tips for improving compliance.

p. ; cm.
ISBN 1-58326-048-X (alk. paper)
1. Pet medicine—United States—Statistics. 2. Pet owners—United States—Statistics. 3. Pets—Services for—United States—Statistics. 4. Veterinary services—United States—Statistics.
[DNLM: 1. Animals, Domestic. 2. Patient Compliance—statistics & numerical data. 3. Data Collection. 4. Veterinary Medicine—ethics. 5. Veterinary Medicine—standards. SF 981 P297 2003] I. American Animal Hospital Association.

SF981.P38 2003
636.089—dc21

2003010061

ISBN 1-58326-048-X

# Contents

# List of Tables

# Understanding the Compliance Gap

Virtually all practice teams are capable of delivering high-quality care to pets. Most likely, you and your team believe you provide the best care possible for your patients. But a comprehensive, in-depth study by AAHA shows that millions of dogs and cats are not in compliance with what we all believe is the best wellness care and treatment of medical problems. (And we studied just the dogs and cats that visit a veterinarian at least once a year!)

What do we mean by compliance? Simply stated, compliance means that the pets in your practice are receiving the care that you believe is best for them. Interestingly, most practice teams think that compliance in their practices is quite high. That may be because most practice teams judge their level of compliance by "gut feel" rather than the results of compliance measurement. In fact, our study showed that in almost every single case, the level of compliance was significantly lower than what practice teams believed.

In this guide, we review the results of the study and quantify the number of pets that receive less than the best care. We provide hard evidence of a substantial problem with compliance, lead you through the steps to improve compliance in your practice, and

demonstrate the financial advantages of increasing compliance. See pages 39 through 84 for a more in-depth description of the study, our methodology, and the results.

Simply stated, compliance means that the pets in your practice are receiving the care that you believe is best for them.

## The Compliance Study

Nonquantitative research completed up to the time of this study suggested that compliance with pet care recommendations was low. Based on that information, we believed that large numbers of pets, even those whose owners regularly took them to see a veterinarian, were not receiving the best possible care. Discussions with various consultants reinforced our suspicions, but empirical evidence did not exist. We determined that more data was needed, and we took our concerns to Hill's Pet Nutrition, Inc. Hill's provided a generous educational grant that allowed AAHA to conduct a comprehensive study of compliance.

We are grateful to Hill's for their support of this project. We believe that the results of the study will provide the profession with extremely useful and actionable information that will ultimately improve the care provided to millions of pets. The commitment of Hill's to all components of high-quality pet care is commendable and greatly appreciated.

The study was composed of six modules:

1. Industry Analysis. In this module, we analyzed existing data available from a wide variety of sources, including

animal health pharmaceutical firms, pet food manufacturers, veterinary laboratories, and large national corporate group practices such as Veterinary Centers of America and Pet's Choice.

2. BENCHMARKING. We reviewed compliance issues in both human medicine and human dentistry to discover parallels or experiences that could benefit the veterinary profession.

3. PRACTICE DIAGNOSTICS. We visited a total of fifty-two practices and/or conducted in-depth interviews with the practice teams. We obtained compliance data in these practices, and observed or discussed with team members their processes, systems, and staff activities related to compliance.

4. ATTITUDINAL AND BEHAVIORAL STUDY. We interviewed veterinarians and practice managers in depth regarding their thoughts and beliefs related to compliance and the steps they felt could be taken to improve compliance.

5. PET OWNER SURVEY. We surveyed over one thousand pet owners regarding the care they provided for their pets, their desires relative to the information and care provided by their veterinarians, and their compliance with health care recommendations.

6. QUANTIFICATION OF THE OPPORTUNITY. We gathered recommendation and compliance

*Our study showed that in almost every single case, the level of compliance was significantly lower than what practice teams believed.*

data from the medical records of almost 1,400 cats and dogs from 240 practices. We then used that data to quantify compliance and the opportunities that practices have to improve pet care by improving compliance.

The study was carried out primarily by Fletcher Spaght, Inc., a Boston consulting firm with prior experience in the veterinary industry. Ethnographic Research, Inc. conducted a portion of the research in the practice diagnostics module, and NFO WorldGroup implemented the pet owner survey. We are also indebted to the insights and advice received from Dr. F. David Schoorman, Professor of Organization Behavior and Human Resource Management at Purdue University's Krannert Graduate School of Management.

## What We Found

The study quantified compliance in six areas:

1. Heartworm testing and preventive
2. Dental prophylaxis
3. Therapeutic diets
4. Senior screenings
5. Canine and feline core vaccines (DHLPP and FVRCP)
6. Preanesthetic testing

Obviously, there are many other areas of companion animal practice in which compliance is also an issue, including routine recheck visits, lab monitoring of patients on medications for chronic conditions, diagnostic procedures, and surgical procedures. While we did not study these other areas in depth, there is *no* evidence that compliance in these areas was any higher than in the six areas we examined.

It is also important to point out that the compliance data reported here *relate only to those dogs and cats that had been seen by a veterinarian at least once in the last twelve months*. According to the AVMA's 2002 *U.S. Pet Ownership & Demographics Sourcebook*, 51 million dogs and 44.2 million cats were in this category (United States only). When we reported the number of pets that were not in compliance with health care recommendations, we *did not include* some 10.6 million dogs and 22.7 million cats in the United States that were not seen by a veterinarian in the past year.

*There were almost 21.5 million dog owners who were either not giving their pets heartworm preventive medication at all or were not giving their dogs the medication for the number of months recommended by their veterinarians.*

## Heartworm Testing and Preventive

There were over 7 million dogs that were not in compliance with their veterinarians' protocol for heartworm testing.

There were almost 21.5 million dog owners who were either not giving their pets heartworm preventive medication at all or were not giving their dogs the medication for the number of months recommended by their veterinarians.

These figures apply only to those dogs in heartworm endemic

Many board-certified veterinary dentists recommend prophylactic treatment for any dog or cat with grade 1 disease or higher.

areas. Areas that are "in transition" (heartworm disease emerging as a threat to the canine population) were not included in these figures. In endemic areas, compliance for testing was 83%, and for preventive medication, compliance was 48%.

In an unrelated study, the American Heartworm Society reported that more than 244,000 dogs tested positive for heartworm in 2001. With the existence of numerous safe and efficacious products for heartworm prevention, there is little justification for almost a quarter of a million dogs becoming infected in just one year.

## Dental Prophylaxis

There were almost 15.5 million dogs and cats with grade 2, 3, or 4 dental disease that had not received dental prophylactic therapy.

Many board-certified veterinary dentists recommend prophylactic treatment for any dog or cat with grade 1 disease or higher. Using that criterion, many millions more pets would have been included in the group that needed dental treatment. Our study revealed 35% compliance for dogs and cats with grade 2 or higher, and only 15% compliance for those with grade 1 disease.

## Therapeutic Diets

The six canine conditions we studied were kidney disease, bladder stones or crystals, food allergies, chronic gastrointestinal disease, acute gastrointestinal disease, and obesity. For cats, we studied the same six conditions plus feline lower urinary tract disease. For these conditions, compliance was 19% for dogs and 18% for cats.

There were 11.6 million dogs with one of six diagnosed conditions that could have been helped by the use of a therapeutic diet that were not fed a therapeutic diet at all or were not fed a therapeutic diet for the appropriate period of time.

More than 9 million cats with one of seven diagnosed conditions that could have been helped by the use of a therapeutic diet were not fed a therapeutic diet at all or were not fed a therapeutic diet for the appropriate period of time.

Compliance with dietary therapy protocols would have been much lower if other conditions known to respond to specific dietary changes were included in the study.

## Senior Screenings

A senior screening includes, at a minimum, blood work and a urinalysis.

There were about 17.9 million dogs considered to be "senior" that had had no diagnostic screens performed in the past year.

There were almost 15.5 million cats considered to be "senior" that had had no diagnostic screens performed in the past year.

Industry data, supported by the data collected in this study, indicated that 35% of dogs and cats in any practice are considered, by most definitions, to be senior. Of those patients, only 32% of dogs and 35% of cats had had diagnostic screening tests performed.

## Core Vaccines

There were 12.4 million dogs and cats that were not protected against core diseases (distemper, hepatitis, leptospirosis, parainfluenza virus, and parvovirus (DHLPP) for dogs and feline viral rhinotracheitis, calicivirus, and panleukopenia (FVRCP) for cats).

While vaccine compliance was higher (87%) than for all other conditions studied, there was still a remarkable number of pets not in compliance with veterinarians' recommended vaccine protocols. It should also be noted that the compliance figure cited may have been artificially high due to the fact that only pets seen within the past year were included in the study.

As with other areas of compliance, we limited our study of vaccines—we reviewed compliance rates only for the core diseases. The addition of other vaccines to the study would no doubt have meant that the compliance rates we measured would have been lower.

## Preanesthetic Testing

Compliance for preanesthetic blood work varied depending on whether the practice required it for all pets, required it only for high-risk patients and strongly recommended it for others, or just recommended it.

Overall, compliance with preanesthetic screening was 72% for dogs and 65% for cats. For those practices that required preanesthetic blood work, compliance was 90%. (The practices allowed clients to sign a waiver, so compliance was not 100%.)

### How Do These Numbers Compare to Your Practice?

It may be hard to believe that there are tens of millions of pets that are not in compliance with basic health care recommendations. Here is a simple chart that will demonstrate the impact on pet health care in just one practice. Table 1 assumes a typical practice with 2.2 full-time-equivalent veterinarians, 1,800 active canine patients, and 1,675 active feline patients. Active patients are defined as patients that have been seen at least once in the last twelve months.

Add your own census data in Table 2 to quantify the number of your patients that are not in compliance with recommended care. (The analysis assumes that your practice's compliance is equal to the national average.)

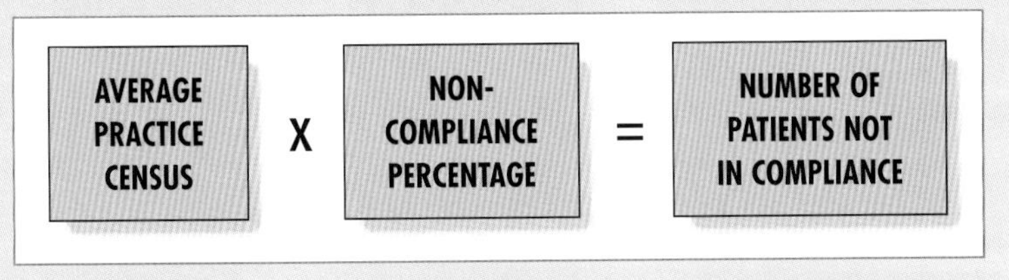

TABLE 1

## Patients Not in Compliance (Average Practice)

| Area of Compliance[1] | Average Practice Census | National Non-Compliance Percentage | Number of Patients Not in Compliance |
|---|---|---|---|
| Heartworm Testing | 1,800 dogs | 17% | 306 |
| Heartworm Preventive | 1,800 dogs | 52% | 936 |
| Dental Prophylaxis[2] | 25% prevalence = 869 dogs and cats | 65% | 565 |
| Canine Therapeutic Diets[3] | 28% prevalence = 504 dogs | 81% | 408 |
| Feline Therapeutic Diets[3] | 25% prevalence = 419 cats | 82% | 344 |
| Canine Senior Screening | 35% senior = 630 dogs | 68% | 428 |
| Feline Senior Screening | 35% senior = 586 cats | 65% | 381 |
| Core Vaccines | 3,475 dogs and cats | 13% | 452 |

NOTE 1: We did not include data for preanesthetic testing because the data varied depending on individual practices' protocols.

NOTE 2: For approximately 19% of the patients studied, there was no grade for dental disease reported in the chart. Based on the ages of these patients, a significant portion of those patients with no grade noted were likely to have had grade 2, 3, or 4 dental disease.

NOTE 3: The rate of prevalence of canine and feline therapeutic diets is only for those seven conditions we studied (see page 9). There are many other conditions that are treatable with therapeutic diets that, if included in the study, would have resulted in even lower compliance rates.

TABLE 2

## Patients Not in Compliance (Your Practice)

| Area of Compliance | Your Practice's Census | National Non-Compliance Percentage | Number of Your Patients Not in Compliance |
|---|---|---|---|
| Heartworm Testing | **All dogs** | 17% | |
| Heartworm Preventive | **All dogs** | 52% | |
| Dental Prophylaxis | **25% of dogs and cats** | 65% | |
| Canine Therapeutic Diets | **28% of dogs** | 81% | |
| Feline Therapeutic Diets | **25% of cats** | 82% | |
| Canine Senior Screening | **35% of dogs** | 68% | |
| Feline Senior Screening | **35% of cats** | 65% | |
| Core Vaccines | **All dogs and cats** | 13% | |

On average, when a veterinarian guesses that compliance is 75%, it is really 50%.

*We found that a major cause of veterinary care providers' failure to make health care recommendations is their misjudgment of the client's willingness to take action.*

findings). Of the 70% of pet owners whose pets had not had screening, 80% indicated that they were unaware of the need and had not been informed by their veterinary care providers.

Of the pet owners who received a recommendation for senior screening, 72% accepted that recommendation. Therefore, we can conclude that the reason why 12 million of the more than 22 million senior pets were not screened was that veterinary care providers failed to make the appropriate recommendation. The failure to make a recommendation was also responsible for a significant portion of compliance failure in the other areas we studied.

**Information overload**

So why do veterinary practice teams fail to make a recommendation for the care they believe is so important? At least some of the problem is attributable to inadequate communication and education. For almost every patient, there is a significant amount of information to convey in a limited amount of time.

In most practices, the staff is already quite busy and does not take the time to carefully make all the necessary recommendations. There are also distractions that preclude delivering recommendations in a way clients can understand. We discuss these issues in more detail on pages 56 and 57.

## Overestimating clients' concerns about money

We found that a major cause of veterinary care providers' failure to make health care recommendations is their misjudgment of the client's willingness to take action. This misjudgment is based on the care provider's perception of either the client's ability to pay or the client's interest in providing the best possible care for her pet.

Every scientifically conducted survey of pet owners since 1995 has demonstrated that the vast majority of pet owners think of their pets as members of the family and are willing to incur the expense of keeping them healthy. Further, these surveys have repeatedly and conclusively reported that the cost of veterinary services is one of the least important factors in choosing a veterinary care provider or in determining satisfaction with veterinary services provided.

These points were again borne out by the pet owner survey that was part of this study. Pet owners were asked to indicate their agreement with one of these two statements:

The cost of the procedure, recommended diet, or other treatment was not a significant factor in compliance.

1. I want my veterinarian to tell me about all of the recommended treatment options for my pet, even if I may not be able to afford them.

2. I want my veterinarian to tell me only about the recommended treatments for my pet that he or she thinks are not too expensive for me.

A full 90% of the respondents chose the first statement, and only 10% chose the second statement. The survey also queried pet owners in order to ascertain the impact that cost had on compliance. The results were as follows:

- Only 7% declined dental care due to cost.
- Only 4% either discontinued or refused therapeutic diets due to cost.
- Only 5% declined senior diagnostic screening due to cost.

While many factors contributed to low compliance, the cost of the procedure or recommended diet or other treatment was not a significant factor in the client's decision to decline care.

Some veterinarians indicated a concern that making multiple recommendations to a client would mean the client would think the veterinarian was motivated by money. However, when we asked pet owners about this issue, 75% agreed or strongly agreed with the statement that their veterinarians made recommendations because they were good for the pet. Only 10% indicated agreement or strong agreement with the statement that veterinarians' recommendations were motivated by a desire to make more money.

The simple step of asking the client to schedule an appointment could significantly increase compliance.

## The Follow-up Component

Assuming that a recommendation has been made, the next factor in the equation is follow-up, or action taken by the practice to ensure that the patient actually receives the care that has been recommended. Follow-up can take many forms, as noted below.

### Scheduling the procedure at the time the recommendation is made

This and previous studies have noted the scenario in which the veterinarian spends significant time explaining the need for a pet to have his teeth cleaned. The explanation generally includes information about the safety and efficacy of the procedure, the benefits to the patient, and a fee estimate. The client leaves the exam room and is provided with an invoice by the receptionist for the fees for that day's services. The client pays the bill and leaves the practice, never having been asked by the receptionist if she would like to schedule an appointment for the dental prophylaxis. The simple step of asking the client to schedule an appointment could significantly increase compliance.

*Almost 80% of pet owners indicated that they wanted instructions verbally and in writing.*

### Providing clear instructions for at-home care and recheck exams

The veterinary health care team routinely gives clients lengthy instructions for at-home care following their pets' medical or surgical care. The instructions often include a recommendation for a follow-up exam. If these instructions are given only verbally, the client may be a victim of information overload. Some instructions, such as the need for a recheck exam, are forgotten. In the pet owner survey component of the study, almost 80% of pet owners indicated that they wanted instructions verbally *and* in writing.

### Sending reminders

Of the six areas we studied, compliance was highest for core vaccines. It is no coincidence that virtually all practices routinely send vaccine reminders. Sending reminders for medication refills and other pet care recommendations is much less common, but a large number of pet owners (65%) indicated a willingness to receive (if necessary) multiple reminders in a variety of ways (e.g., phone, mail, email). Seventy-two percent said they would like to receive a phone call if they are overdue for a recommended treatment or preventive measure.

Based on this information, we can conclude that veterinary care providers could substantially improve compliance with recommendations in all areas by sending reminders. For example, you can improve compliance for heartworm preventive by sending timely reminders to those clients who purchase less preventive than was recommended.

### Making follow-up phone calls

In the pet owner survey, 78% of respondents indicated that they would like to receive a follow-up phone call from the practice after a visit. However, only 52% of the respondents reported that they had ever received such a call! Follow-up calls are particularly helpful when therapeutic diets are dispensed and clients are making the transition to a new food. Those practices that consistently followed up on therapeutic diets by telephone achieved significantly higher client compliance. Follow-up calls also greatly increased compliance with follow-up visits and exams.

*Of the three components of the compliance equation, client acceptance appears to be the smallest factor.*

The pet owner survey revealed that of those pet owners who did not return for a recheck exam as directed by their veterinarians, 38% would have done so if they had received appropriate follow-up from the practice. While these data do not indicate that follow-up would result in 100% compliance, it would nonetheless represent substantial improvement over today's typical level of compliance with recommendations for rechecks.

## The Client Acceptance Component

If the veterinarian or other team member makes a recommendation and follows up on that recommendation, the final decision to comply is ultimately the pet owner's. As noted on page 18, a small percentage of clients will not comply due to cost. Time constraints and inconvenience also play a small part in some clients' failure to take action. Further, while some clients have good intentions, they simply fail to comply. In these cases, the practice team was not able to convince the owner of the need for or benefit of the recommended service.

If you fail to detect a treatable condition in a geriatric patient because you did not recommend a senior screening, can you claim that you practice high-quality medicine?

However, of the three components of the compliance equation, client acceptance appears to be the smallest factor. In fact, *there was nothing in this study that would suggest that pet owners are the primary barrier to compliance*. Veterinarians and their staff have much more influence than they believe and could have a substantial impact on improving compliance.

## Why Should You Try to Improve Compliance?

Compliance is clearly a quality-of-care issue. If you, as a veterinary care professional, have the knowledge, skills, equipment, medicines, and diets available to maintain wellness and restore health when needed, and you fail to recommend appropriate care and take reasonable steps to ensure client acceptance of those recommendations, you have not achieved the highest-quality *outcomes* possible for your patients.

- If you fail to recommend a therapeutic diet for a dog with kidney disease and thereby fail to help the client achieve compliance with that diet, have you truly done all that you can to extend the life of that patient?

- If you fail to detect a treatable condition in a geriatric patient because you did not recommend a senior screening, can you claim that you practice high-quality medicine?

- If, as noted on page 8, more than 244,000 active patients test positive for heartworm disease every year, do you think you are doing your part?

"Outcomes assessment" is a focus of patient care in human medicine, as well it should be. While it is obviously important, for example, for a skilled cardiologist to get a patient through an acute episode of coronary artery disease, the length and quality of that patient's remaining life is equally important. In the benchmarking portion of the study, we found that several cardiac care facilities were measured and judged on long-term patient survival rates, which were directly related to patient compliance.

The top-rated cardiac care units were those whose staff took an active role in ensuring their patients' compliance with recommendations for medications, diet, and lifestyle changes. If you want to be known for the quality of care you provide, assess your own patient outcomes and the role that low compliance plays in those outcomes.

## The Cost-Benefit Analysis

The study demonstrates that you can improve compliance by decreasing the "recommendation gap" and by increasing follow-up with clients. Shrinking the recommendation gap is simply a matter of requiring every staff member to commit to making the recommendations that represent the best possible care for every single patient. The practice does not incur any additional cost to make the *recommendation.*

Improving follow-up does involve additional staff time, and perhaps some additional expense, such as postage. But as you can see in Table 3, increasing compliance by just ten percentage points for a typical practice (2.2 full-time-equivalent veterinarians and 3,475 active patients) results in 1,287 additional treatments provided to patients, $132,535 in additional revenue, and an additional $81,364 in gross profit.

*For a typical practice, increasing compliance by just ten percentage points results in 1,287 additional treatments provided to patients, $132,535 in additional revenue, and an additional $81,362 in gross profit.*

Although many veterinary practice consultants and the KPMG Megastudy (*The Current and Future Market for Veterinarians and Veterinary Medical*

TABLE 3

## How a 10% Increase in Compliance Affects Revenue and Gross Profit[1,2]

| Compliance Area | Current Overall Rate of Compliance | Compliance Goal | Number of Additional Treatments per Year | Increase in Revenue Through Improved Compliance | Increase in Gross Profit Through Improved Compliance[3] |
|---|---|---|---|---|---|
| Dental Prophylaxis | 29% | 39% | 158 | $31,521 | $29,945 |
| Core Vaccines and Related Exams | 87% | 97% | 337 | $16,294 | $15,316 |
| Therapeutic Diets | 5% | 15% | 202 | $48,587 | $12,633 |
| Senior Screenings | 33% | 43% | 119 | $13,483 | $9,843 |
| Heartworm Preventive | 48% | 58% | 180 | $11,988 | $5,994 |
| Preanesthetic Testing | 63% | 73% | 111 | $6,072 | $4,190 |
| Heartworm Testing | 83% | 93% | 180 | $4,590 | $3,443 |
| Total Improvement | | | 1,287 | $132,535 | $81,364 |

NOTE 1: Calculations include only the areas we studied, with the exception of diets, which, for the quantification of the opportunity, we expanded beyond the diagnoses studied.

NOTE 2: Revenue and gross profit were calculated using data compiled from a variety of sources, including *The Veterinary Fee Reference*.

NOTE 3: Gross profit is defined as revenue less the direct costs of providing the care, including drugs and supplies, but not staff time.

*Services in the United States*) suggest that most practices have the capacity to increase services by 10–20% without adding staff, let's assume that additional staff time is needed for proper follow-up. Assume that increasing compliance in the seven areas in Table 3 by ten percentage points will require hiring an additional staff position at an annual cost of $26,000 ($10 per hour wage plus 25% benefits).

We concluded that providing the additional care would generate $132,535 of additional revenue for a typical practice, which equates to a gross profit of $81,364. If you subtract the cost of the additional staff person, you still gain an additional gross profit of $55,364. This money can be invested in additional equipment, more staff training, continuing education, higher wages, or more benefits.

Most practices have the capacity to increase services by 10–20% without adding staff.

# Six Steps to Improving Compliance and Patient Care

In the first section of this book, "Understanding the Compliance Gap," we proved that a tremendous number of dogs and cats are not in compliance with what we all believe to be the best protocols for health care. Low compliance equates to less-than-the-best patient care and lower-than-possible revenue and profits. In this section, we demonstrate how you can take advantage of the opportunity to increase compliance. The steps on pages 29 through 38 lead you through a process for improvement, starting with measuring current compliance and ending with celebrating successes. Read on to learn what you can to do ensure that your patients receive the very best care possible.

## Step 1: Measure Current Compliance

While we identified a few practices that had above-average compliance rates in one or two areas, we found no practices that had high rates of compliance in all, or even most, areas. The first step on the road to improving compliance in your practice is finding out where you stand and which areas are most in need of improvement.

If you have visited AAHA's traveling exhibit, AAHA! Driving Excellence in Veterinary Practice℠, you received a CD-ROM containing a tool to help you measure compliance through a limited chart review of your active patients. Or you may have received a CD that contains the tool along with this book. (If you have misplaced your compliance measurement CD, call the AAHA Member Service Center at 800/883-6301 to order a replacement.)

*The first step on the road to improving compliance in your practice is finding out where you stand and which areas are most in need of improvement.*

The chart review contained in the tool, which should take a technician or receptionist no more than two hours to complete, will provide you with a baseline compliance percentage for your practice. The tool allows you to establish the compliance parameters you believe are appropriate in order for your patients to receive optimal health. That is, you will use your own protocols, not those established by others.

There really are no shortcuts to measuring compliance in your practice. Completing a review of a random sample of charts is the only way to obtain baseline numbers for compliance in most areas. The two hours the staff spends entering data from the charts and the time you spend reviewing the results and deciding on your protocols may be the best investment in your practice you will ever make. Making a conscious decision to improve patient care and then measuring that level of care will make a difference for hundreds, if not thousands, of your patients.

## Step 2: Involve the Entire Staff

Our study clearly showed that practices that involved the staff in compliance issues achieved significantly higher levels of compliance than those that did not. Staff involvement is a vital component of making wellness recommendations, reinforcing recommendations made by the veterinarian, providing client education, and following up by phone. Staff members need to be trained on and committed to clinic protocols. Here are specific steps to take to involve the entire staff in increasing compliance:

- Assemble the staff and establish protocols that everyone can agree to follow. There are existing guidelines for heartworm testing and prevention (American Heartworm Society at *www.heartwormsociety.org*), feline vaccinations (American Association of Feline Practitioners at *www.aafponline.org*), and canine vaccinations (AAHA at *www.aahanet.org*) that you can use as is or as a starting point for establishing practice-specific protocols. Establish protocols for services such as senior profiles, preanesthetic screening, dietary management, dental prophylaxis, and follow-ups for pets on medications for chronic conditions. Be specific, and commit the protocols to writing. Disseminate the protocols to all staff members, and make them a part of orientation and training for all new staff members.

*Our study clearly showed that practices that involved the staff in compliance issues achieved significantly higher levels of compliance than those that did not.*

Creating protocols everyone can agree to follow is a critical step, so don't rush it. You can accomplish this goal in a series of meetings with staff over time, or you can assign small teams to draft protocols for certain compliance areas. The entire group can then review the drafts.

Ensuring that your staff is committed to your protocols is very important. Staff may undermine compliance efforts if they do not believe that the protocols are truly in the best interest of the patients in your care.

- Next, close the "recommendation gap." This means getting everyone, particularly the doctors, to agree that they will *always* make a recommendation consistent with the practice's protocols *without prejudging the client's level of interest in providing the best care for her pet or her willingness to pay*. You and the entire staff must assume that clients will accept your recommendations—all you have to do is ask.

- Determine the specific follow-up steps that the practice will take, and assign responsibility for each task. For the highest success rates with follow-up, implement these procedures:

  — Assign someone to pull and review patient records prior to the client's arrival. That person will review the pet's compliance status for all products and services the practice team has identified as important. That person will also make notations for areas in which the pet is out of compliance or that require follow-up. Staff should be mindful of clues to noncompliance, such as the date and amount of past heartworm preventive purchases. Upon the client's arrival, the receptionist can say, "I see that

Spike is due for his annual heartworm test. I can have our veterinary technician draw the blood sample before you see the doctor so that you'll have the results before you leave."

Assign someone to pull and review patient records prior to the client's arrival. That person will review the pet's compliance status for all products and services the practice team has identified as important.

— Keep educational materials and resources close at hand. Use brochures available from AAHA, pet food companies, or animal health manufacturers. You can also create your own. Recommend websites, such as AAHA's *www.healthypet.com* and Hill's *www.hillspet.com*, that clients can visit for reliable and accurate information. Use these materials to reinforce the recommendations you've made for the pet's care.

— Schedule the next appointment or recommended procedure before the client leaves the practice. This can be as straightforward as the receptionist saying to the client at discharge, "I see that the doctor has recommended that Fluffy return for a recheck exam in two weeks. Would this same time on the 25th be good for you?"

— Develop a system for providing written instructions for follow-up and at-home care that will reinforce and encourage compliance with verbal instructions. Some practice management software includes a module with written handouts, and there are other resources, such as

The steps of measuring and tracking results are critical to success. If you are unable to use your practice management system to track compliance, consider a manual, paper-based system or a hybrid of your software system supplemented by paper records. A paper system can be as easy as making a log on a piece of paper. If one of your goals is to increase the percentage of pets that receive senior screening, the client relations specialist can simply write down the name of every senior pet that comes in, followed by a notation of "Yes," "No," or "Schedule follow-up phone call."

## Step 6: Celebrate Victories!

As with all goal-setting strategies, it is important to celebrate successes. Set milestones, and celebrate as you achieve each milestone rather than waiting until you have achieved the overall goal. For example, let's say your goal is to increase senior screening compliance from 35% to 45% in twelve months. You could celebrate with a pizza lunch or an afternoon ice cream break for the entire staff when compliance reaches 40%. You could also create a large "thermometer" chart and color in the tube as compliance increases, celebrating at predetermined intervals along the way.

Remember to measure compliance in terms of the number of pets receiving better care. For example, tell your staff, "As a result of our efforts to increase compliance for senior screening, fourteen senior dogs and eleven senior cats were screened this month. We found problems in three of these patients that we can now treat before they become more serious." Measuring progress toward the goal and celebrating successes along the way will help staff maintain their commitment to improved compliance.

# Executive Summary of *Compliance in Companion Animal Practices*

## Introduction

This Executive Summary provides complete results of the study in significantly greater detail than in the section entitled "Understanding the Compliance Gap" on pages 1 through 25. It also includes, where appropriate, the study methodology.

Nonquantitative research completed up to the time of this study suggested that compliance with pet care recommendations was low. Based on that information, we believed that large numbers of pets, even those whose owners regularly took them to see a veterinarian, were not receiving the best possible care. Discussions with various consultants reinforced our suspicions, but empirical evidence did not exist. We determined that more data was needed, and we took our concerns to Hill's Pet Nutrition, Inc. Hill's provided a generous educational grant that allowed AAHA to conduct a comprehensive study of compliance.

We are grateful to Hill's for their support of this project. We believe that the results of the study will provide the profession with extremely useful and actionable information that will ultimately improve the care provided to millions of pets. The commitment of Hill's to all components of high-quality pet care is commendable and greatly appreciated.

There were six primary areas of focus for this study: heartworm testing and preventive, dental prophylaxis, therapeutic diets, senior screening, canine and feline core vaccines (DHLPP and FVRCP), and preanesthetic testing. The study was composed of six modules as described below.

1. Industry Analysis. In this module, we analyzed existing data available from a wide variety of sources, including animal health pharmaceutical firms, pet food manufacturers, veterinary laboratories, and large national corporate group practices such as Veterinary Centers of America and Pet's Choice.

2. Benchmarking. We reviewed compliance issues in both human medicine and human dentistry to discover parallels or experiences that could benefit the veterinary profession.

3. Practice Diagnostics. We visited a total of fifty-two practices and/or conducted in-depth interviews with the practice teams. We obtained compliance data in these practices, and observed or discussed with team members their processes, systems, and staff activities related to compliance.

4. Attitudinal and Behavioral Study. We interviewed veterinarians and practice managers in depth regarding their thoughts and beliefs related to compliance and the steps they felt could be taken to improve compliance.

5. Pet Owner Survey. We surveyed over one thousand pet owners regarding the care they provided for their pets, their desires regarding the information and care provided by their

veterinarians, and their compliance with health care recommendations.

6. QUANTIFICATION OF THE OPPORTUNITY. We gathered recommendation and compliance data from the medical records of almost 1,400 cats and dogs from 240 practices. We then used that data to quantify compliance and the opportunities that practices have to improve pet care by improving compliance. We also studied the practice's protocols for each compliance area (e.g., number of months of heartworm preventive recommended, age at which pets are considered "senior").

The study was carried out primarily by Fletcher Spaght, Inc., a Boston consulting firm with prior experience in the veterinary industry. Ethnographic Research, Inc. conducted a portion of the research in the practice diagnostics module, and NFO WorldGroup implemented the pet owner survey. We are also indebted to the insights and advice received from Dr. F. David Schoorman, Professor of Organization Behavior and Human Resource Management at Purdue University's Krannert Graduate School of Management.

## Module I: Industry Analysis

The purpose of this module was to formulate a preliminary understanding of compliance by analyzing the results of veterinary suppliers' existing research. We obtained a great deal of information from previous studies and the marketing data from a number of companies.

Only 44% of cats are vaccinated with core vaccines.

A total of ten pet food companies, pharmaceutical manufacturers, and laboratory test producers provided meaningful data and input. In addition, three of the large national corporate groups shared their experiences, as did the Feline Health Center at Cornell University, the American Association of Feline Practitioners, a large national veterinary laboratory, and several independent veterinary consultants. The result of this portion of the study was the calculation of the compliance gap and potential for improvement in compliance levels based on existing data (e.g., the number of dogs and cats that visit a veterinarian annually).

In addition to the six primary areas of focus, we also gathered compliance data in the areas of allergy testing and treatment, feline immunodeficiency virus (FIV) testing, keratoconjunctivitis sicca (dry eye) treatment, and implantable microchips. The assumptions we used in calculating compliance gaps are detailed in Table 4.

TABLE 4
## Assumptions Used in Calculating Compliance Gaps

| | | |
|---|---|---|
| Dog Population | 63,700,000 | Average of the year 2000 data from the American Pet Products Manufacturers Association (68M) and the Pet Food Institute (59.4M) |
| Dogs Receiving Veterinary Care | 54,300,000 | 85.3% of dog-owning households use veterinary care (AVMA) |
| Cat Population | 74,000,000 | Average of the year 2000 data from the American Pet Products Manufacturers Association (73M) and the Pet Food Institute (75.1M) |
| Cats Receiving Veterinary Care | 50,100,000 | 67.7% of cat-owning households use veterinary care (AVMA) |
| Small-Animal Full-Time-Equivalent Veterinarians | 34,261 | Derived from 2001 AMVA data:<br>100% of small-animal-exclusive veterinarians<br>75% of small-animal-predominant veterinarians<br>50% of mixed-animal veterinarians<br>25% of large-animal-predominant veterinarians |

Just 34% of dogs in heartworm endemic states are in compliance with heartworm preventive recommendations.

## Estimates of Compliance Gaps

Analysis of the data available from all sources resulted in the following estimates of compliance gaps as presented in Table 5.

TABLE 5

### Estimates of Compliance Gaps[1,2]

**Allergy Testing**

Additional revenue potential per veterinarian per year ................................ $1,900

Prevalence of atopy in dogs: 10%

Compliance: Approximately 20%

**Allergy Treatment**

Additional revenue potential per veterinarian per year .............................. $18,900

Prevalence of atopy in dogs: 10%
(of which 60% are responsive to immunotherapy)

Compliance: 18%

**Canine Core Vaccines**

Additional revenue potential per veterinarian per year .............................. $12,600
(does not include fees for related examinations)

Percentage eligible for vaccination: 100%

Compliance: 56%

**Feline Core Vaccines**

Additional revenue potential per veterinarian per year .............................. $12,900
(does not include fees for related examinations)

Percentage eligible for vaccination: 100%

Compliance: 44%

**Dental Prophylaxis**

Additional revenue potential per veterinarian per year............................ $310,000

Prevalence of periodontal disease: 85% of dogs and cats over one year of age

Compliance: 3% for dogs; 1% for cats

**FIV and FeLV Testing**

Additional revenue potential per veterinarian per year................................ $9,100

Prevalence: 25% at risk and/or showing signs of disease

Compliance: 24%

**Heartworm Testing (Canine)**
Additional revenue potential per veterinarian per year
(for veterinarians practicing in endemic states only)
If testing annually ........................................ $26,400
If testing every two years ........................................ $5,400
Percentage of canine population residing in heartworm endemic states: 81%
Compliance: 37% if all were tested annually;
74% if all were tested every two years

**Heartworm Preventive (Canine)**
Additional revenue potential per veterinarian per year
(for veterinarians practicing in endemic states only) .................... $44,000
Percentage of canine population residing in heartworm endemic states: 81%
Compliance: 34%

**Senior Screening**
Additional revenue potential per veterinarian per year .................... $114,600
Percentage considered senior: 35% of all dogs and cats
Compliance: 14%

**Therapeutic Diets**
Additional revenue potential per veterinarian per year .................... $110,300
Percentage of all dogs and cats that have visited a veterinarian that had a diagnosis that would benefit from treatment with a therapeutic diet: 59%
Compliance: 12%

**TOTAL ADDITIONAL REVENUE OPPORTUNITY**
**per veterinarian per year .................... $639,700 – $660,700**

NOTE 1: The compliance rate is the percentage of pets eligible for treatment that were actually in compliance with the recommendation.

NOTE 2: Revenue opportunity was calculated using the assumptions in Table 4 and average fees compiled from a variety of sources, including *The Veterinary Fee Reference*.

It should be noted that the calculation of additional revenue opportunity is based on the specific products and services we have listed and does not include the revenue that could be generated by increased compliance with recheck exams, with lab work for patients on medications for chronic conditions, and in other areas of poor compliance. The revenue projection also omits revenue that could be earned from additional exams and treatment of conditions diagnosed by screenings.

*The national groups reported that quality of care motivates staff to improve compliance more than increased revenue for the practice does.*

## National Corporate Veterinary Groups

The national corporate veterinary groups that contributed to this study were aware of low compliance. While these groups had the size and resources that enabled them to improve compliance, most lacked the infrastructure necessary to accurately track meaningful compliance data (though all reported that they were building this infrastructure). All of the groups reported that they were seeing significant improvement as a result of monitoring compliance.

In general, the corporate groups had more consistent and a higher level of staff training, and they recognized the important role that staff plays in achieving compliance and client satisfaction. Further, the national groups reported that it was important to maintain a strong focus on medical and quality-of-care benefits rather than on the economic benefits accruing from improved compliance. Quality of care motivates staff to improve compliance more than increased revenue for the practice does.

## Characteristics of Practices Achieving Higher Levels of Compliance

We asked the consultants and organizations that participated in this part of the study to comment on the characteristics of practices that achieve higher-than-average compliance. Here are the most common characteristics:

1. Practice management. Practices with high compliance are generally well-managed, efficient operations that track the use of tests and treatments to diagnose and treat health problems. Most have a business plan in place, and most benchmark their performance against previous years and against other practices.

2. Processes. Those practices that achieve higher compliance are thought to have better systems and processes, including clear protocols for making appointments upon discharge and use of an efficient and effective reminder system. Every staff member clearly understands all protocols and makes appropriate recommendations to clients based on those protocols.

> *Those practices that achieve higher compliance are thought to have clear protocols for making appointments upon discharge and use of an efficient and effective reminder system.*

3. Veterinarians' behavior. Doctors in practices with high levels of compliance present all alternatives to clients rather than prejudging a client's interest or ability to pay. These doctors deliver their opinions with confidence and provide a high level of client education.

4. PRACTICE STAFF. These practices also tend to have a higher technician-to-veterinarian ratio (3 to 1 as compared to the national average of 1.8 to 1 reported in AAHA's *Financial and Productivity Pulsepoints*, Second Edition) and have well-trained and well-paid staff (resulting in lower turnover rates). Well-trained practice team members reinforce the doctors' messages and reduce the doctors' burden of being solely responsible for recommendations and follow-up.

## Module II: Benchmarking

We reviewed existing literature on compliance in human medicine and dentistry, and we interviewed eleven physicians and dentists at leading universities and medical centers. In addition, we interviewed officials at nine national associations and state public health departments, along with executives at three pharmaceutical companies and managed-care organizations.

### Human Dentistry

We did not gain new information from our review of compliance rates in human dentistry. Demand for services in human dentistry is increasing fairly dramatically because the profession serves a growing population that is living longer and keeping more teeth. As one dental executive put it, "There has been a significant increase in the number of teeth requiring care." In addition, there has been a shift in demand for services that deal with aesthetics, such as whitening, implants, and adult orthodontia.

While demand for services in human dentistry has increased dramatically, the number of dentists per 100,000 people has been decreasing since 1990 and is projected to continue to decrease for many years. The combination of increasing demand and decreasing supply has had a very positive economic impact on dentists.

In addition, we found that compliance was not a great concern for human dentists. Dental insurance plans now cover an estimated 145 million people in the United States, or approximately 50% of the total population. Dental insurance plans are largely prepaid plans that cover wellness procedures, providing an incentive for a large number of people to obtain the routine care recommended by the profession.

Also, adults who seek out dental services for aesthetic reasons are motivated to comply with dental recommendations. Low compliance with at-home care and follow-up visits is not an issue with these patients.

## Human Medicine

Compliance in human medicine depends on two factors:

1. How often the practitioner recommends a clinical intervention for the right patient at the right time

2. Once the clinical intervention is recommended, how often the patient follows through

There are many examples of compliance gaps in human medicine. While annual cholesterol tests and eye exams for diabetics are strongly recommended by the American Diabetes Association, only 46% of diabetics have their cholesterol checked annually, and only 29% have an annual eye exam, according to a study of 15,893 diabetic patients conducted by The Medstat Group.

*Compliance is not a great concern for human dentists. Dental insurance plans cover wellness procedures, providing an incentive for a large number of people to obtain the routine care recommended by the profession.*

The treatment of coronary artery disease (CAD) is a well-studied example of poor compliance in human medicine. Heart disease is the leading cause of death in the United States. Patients with known CAD are five to seven times more likely to experience a cardiac event than the general population. Clinical trials and other studies have demonstrated that the risk of recurrent events can be reduced by 20–30% with the use of each of three medications: aspirin, beta blockers, and ACE inhibitors. However, in one study of 186,000 patients with known CAD, compliance rates were only 86% for aspirin, 50% for beta blockers, and 59% for ACE inhibitors. The American Heart Association estimates that 80,000 lives could be saved annually if hospitals implemented its published guidelines for managing risk factors in patients with cardiovascular disease just 85% of the time.

Two specific programs in human medicine have been able to significantly increase compliance, and we describe each of those programs below.

### Cardiovascular Hospitalization Atherosclerosis Management Program (CHAMP) at UCLA

The CHAMP program uses a multimodal approach to improve

compliance, but much of their improvement is due to the fact that they made compliance a measurement of success. That is, patient outcomes have become an important measure, and physicians and other medical professionals have recognized their roles and responsibilities in improving patient outcomes. The CHAMP program at UCLA was able to increase prescription rates for statins, ACE inhibitors, and beta blockers for discharged acute myocardial infarction patients from 6–12% to 56–58% in just two years.

### The Center for Stroke Research in Chicago

This center has achieved significant increases in patient compliance through a concerted effort to improve patient education and to increase collaborative efforts between physicians, nurses, and pharmacists. A critical component was follow-up with patients by pharmacists.

*Patient outcomes have become an important measure of success, and physicians and other medical professionals have recognized their roles and responsibilities in improving patient outcomes.*

There were a number of findings regarding compliance in human medicine that have implications for veterinary medicine.

- GUIDELINE-BASED, QUALITY-IMPROVEMENT INITIATIVES. The existence of clear, evidence-based guidelines (e.g., how often and at what age women should have mammograms) helps provide a framework for care and doctor recommendations. Guidelines can also increase doctors' confidence in making recommendations. Measuring adherence to recommended guidelines is very important.

- Practitioner education programs sponsored by pharmaceutical companies. Investment in educational programs directed toward veterinarians can help improve doctor compliance with accepted guidelines and quality-of-care initiatives. Symposia on targeted diseases that can be managed with the proper use of pharmaceuticals or therapeutic diets are excellent ways to increase awareness.

- Accreditation standards that include collection and analysis of performance measures. Quality-of-care measures, when included as a part of accreditation programs, can significantly increase adherence to care guidelines and can provide an additional incentive to address compliance issues. Practices that actually measure compliance are much more likely to achieve higher compliance. It should be noted that the Joint Commission on Accreditation of Healthcare Organizations (JCAHO—the accrediting body that covers 95% of hospital beds in the United States, as well as managed care organizations and nursing homes) has recently started requiring monitoring of outcomes and compliance at accredited institutions.

- Consumer advocacy groups that endorse guidelines and increase public awareness. The endorsement of guidelines by such groups as the American Heart Association, the American Cancer Society, and the National Organization for Women has improved physician and patient compliance in human medicine. Endorsement of guidelines in veterinary medicine by such groups as AAHA, AVMA, species groups, humane organizations, and the American Kennel Club can have a positive impact on pet owners' awareness of the need for proper care.

- Patient/client education and excellent patient/client-doctor relationships that enhance compliance. Taking time to establish positive relationships with patients and clients fosters increased compliance. Similarly, good relationships generally result in higher levels of client education, which further increases compliance.

# Module III: Practice Diagnostics

There were two components of this module. One component involved a team of ethnographers that visited fourteen veterinary practices (West Coast, Southeast, and Midwest locations) to observe doctors' and staffs' relationships with clients, as well as their behaviors related to compliance. The second component involved on-site visits or in-depth interviews with the doctors and staff of thirty-eight practices across the country to gather compliance data and review processes and systems in place for measuring compliance, making recommendations, and following up with clients.

## Ethnographic Observations

The team of ethnographers videotaped numerous interactions in the veterinary practices they visited. The videotaped sessions provided vivid illustrations of a number of interesting findings related to compliance.

### Veterinarians' desire to please the client

Other studies have suggested that veterinarians have a strong desire to be liked by their clients, and more specifically, that they fear client rejection. This fear of negative client reaction was

The existence of clear, evidence-based guidelines can increase doctors' confidence in making recommendations.

cited in one study as one of the reasons that the veterinarians who were surveyed did not always recommend the best possible treatment.

One videotaped session involved a veterinarian examining a dog with an eye problem. He dispensed ophthalmic ointment and instructed the owner to apply the ointment to the eye four times each day. The owner, obviously deeply attached to her dog (she repeatedly referred to the pet as "my baby") reacted negatively to the instructions, fearing that she would upset her pet. The veterinarian immediately backed off from his instructions, saying, "Well, try to apply it at least once per day." In so doing, he directly undermined compliance with the recommended treatment.

**Presentation of recommendations and large volumes of information**

A scene from a busy veterinary practice demonstrated that the need to impart a large amount of information to a client in a limited amount of time has a negative impact on compliance. A veterinarian is shown continuing her conversation with her client at the discharge window. It is obvious that she felt a need to get on to the next patient in her very busy practice. She had discussed a wide range of recommendations with the client, including diet, heartworm prevention, dentistry, and medication for an ongoing skin condition. The client clearly had not absorbed all of the information and recommendations, and the doctor ended the conversation at a point at which the client was unlikely to

and did not account for underdiagnosis.) There was a notable "recommendation gap," often due to the veterinarian's belief that a pet may not accept a dietary change, or that a client would be unwilling to pay the higher cost of a therapeutic diet.

In addition, obesity in pets was significantly underdiagnosed. Overall, medical record data showed that a diagnosis of obesity was made for only 13% of cats. (If we include just those cats over fifteen pounds in the definition of "obese," we calculate an obesity rate of 17%. If we include those cats that weigh more than twelve pounds, the obesity rate increases to 25%.) The same chart review indicated that veterinarians did not make a diagnosis of obesity for a significant number of cats weighing more than twelve pounds. In fact, in a significant number of cases, veterinarians did not make a diagnosis of obesity even for cats weighing more than fifteen pounds.

*Those clients who were given feeding tips and "moral support" were able to successfully transition a pet to a new diet, whereas those that did not receive follow-up were not as successful.*

Overall compliance with therapeutic diets was 18% for cats and 19% for dogs. The compliance gap was the product of under-recommendation, no purchase of product by the client, and the feeding of additional products by those clients who did purchase therapeutic diets. If we added the pets for which no diagnosis of obesity was made, the compliance rates we calculated would have been even lower.

This component of the study clearly demonstrated the value of follow-up by the practice. Those practices that achieved the highest compliance with dietary recommendations made consistent use of staff to follow up with clients. It was clear that many clients needed help in transitioning a pet from his regular food to a therapeutic diet. Those clients who were given feeding tips and "moral support" were able to successfully transition a pet to a new diet, whereas those that did not receive follow-up were not as successful. Practices that consistently followed up with clients whose pets were on a new diet reported a much higher percentage of patients staying on the new diet and much higher compliance with recommended feeding guidelines.

### Compliance for other services studied

The compliance data obtained from this module was combined with data obtained in Module V and is reported on pages 66 through 76. The practices in this portion of the study, like those in Module V, consistently overestimated their compliance levels.

## Communication and Behavior as Contributors to Compliance

The practice team members that participated in this portion of the study exhibited a wide spectrum of communication and other behaviors that influenced compliance. It was quite clear that those practices that used their staff to make or support recommendations had higher compliance rates. Leveraging staff generally resulted in repetition of the message to the client, and repetition improved compliance.

Other aspects of staff involvement that affected compliance included the amount of time staff members were able to spend with clients, the level of language they used with clients, and their rapport with clients. The degree of staff training varied considerably among the participating practices. Training programs ranged from excellent, with emphasis on client communications through role playing, to poor or nonexistent.

Many practices provided client educational materials in the form of written instructions and recommendations as well as in the form of practice- and commercially generated educational materials. (These practices cited a preference for materials they generated themselves but acknowledged use of commercial materials due to their convenience.)

Leveraging staff generally resulted in repetition of the message to the client, and repetition improved compliance.

### Veterinarians' views of compliance barriers

Veterinarians overwhelmingly cited cost and insufficient client communication/education as the primary barriers to compliance. Cost was mentioned as a cause of noncompliance by 65% of the participants; communication and client education were mentioned by 57% of the participants. Lack of follow-up by the veterinarian was cited by 10%, and poorly educated staff was cited by 5%.

Veterinarians saw education and communication as being key factors to improving compliance. Staff education ranked highest, followed closely by client education. Improved communication was ranked as the next most-needed factor to improve compliance, followed by better processes and systems.

When making recommendations, veterinarians overwhelmingly anticipated that cost would be the clients' primary objection. (As will be noted in the section on pet owner research, this fear has been blown out of proportion.) They cited convenience and the clients' fear of safety for their pets as much less likely client objections.

A small but noteworthy group believed that compliance was strictly the responsibility of the client. This group believed that the practice's only responsibility was to provide the information one time, and after that, "It is up to the client." In Module IV, as noted on page 65, it was evident that a much larger group placed a significant amount, but not all, of the responsibility on the client.

## Module IV: Behavioral and Attitudinal Research

This module consisted of sixty-seven in-depth interviews with AAHA-accredited members, nonaccredited AAHA members, and nonmembers. The sixty-seven practices were from all regions of the country and ranged in size from one to eight full-time equivalent veterinarians (average 2.9). Revenue for the practices in this group ranged from $100,000 to $3,200,000 (averaging $980,000).

### Key Findings

Most of the veterinarians we interviewed saw compliance as a quality-of-care issue, although most recognized the relationship between quality medicine and increased revenue. They believed that staff would be more motivated by a patient-care message than an economic message.

Over half of the veterinarians we interviewed were satisfied with their current compliance levels because they believed that their compliance numbers were high. They acknowledged that the belief that their numbers were high was predominantly based on "feel" versus true compliance monitoring.

A significantly large group of practice owners purported to be happy with compliance because they accepted a degree of noncompliance. These veterinarians believed that their current patient-education and reminder systems were adequate and that noncompliance was solely or partly the clients' responsibility. More than twice as many participants believed that compliance was the client's responsibility than believed it was the veterinarian's responsibility (shared responsibility fell in between the two).

Most of the veterinarians we interviewed accepted the adage that "What gets measured gets managed," and believed that monitoring compliance would be an effective tool for improving compliance. However, many expressed a fear that compliance monitoring would overburden staff.

Most practices had computer systems and would expect to use them as a critical aid in compliance monitoring. However, when pushed on details about their systems, veterinarians admitted to not being very knowledgeable. Without built-in tracking and reporting capabilities in their software systems, many practices may not be technologically able to pursue monitoring and tracking of compliance.

About 25% of the veterinarians we interviewed did not involve staff at all in delivering wellness recommendations, primarily due to training issues and staff turnover. Another 30% did not involve staff to any significant extent.

*Over half of the veterinarians we interviewed were satisfied with their current compliance levels because they believed that their compliance numbers were high.*

There was wide support among participants for AAHA to introduce compliance "guidelines" to the profession. Although somewhat skeptical that AAHA guidelines would have a significant impact on clients, most believed that the guidelines would help convince staff of the need to make appropriate health care recommendations to clients.

## Module V: Quantification of the Opportunity

This module consisted of gathering medical-record data from a large number of practices in an attempt to measure compliance and quantify the opportunity for improved patient care and increased practice revenues. We also asked questions related to the practice's demographics; pet health care protocols; compliance monitoring; client communications; and staff involvement in compliance, care recommendations, and client follow-up.

The survey was conducted via the Internet over a four-week period late in 2002. (Participants were also given the option of printing the survey and returning the completed form via fax.) Participants included nonrandomly selected accredited and nonaccredited AAHA members and nonrandomly selected Hill's Pet Nutrition customers, which included AAHA members and nonmembers.

Doctors from 242 practices participated in this portion of the study. The participating practices averaged 2.55 full-time equivalent veterinarians. About half of the practices' revenue fell in the range of $700,000 to $850,000. Practices were provided with instructions for randomly selecting the medical records of three dogs and three cats for review and were instructed that the patients whose records were reviewed should not be from the same family.

It is important to note that participants were instructed to select the medical records only of pets that had visited the practice in the past twelve months. We chose this definition rather than "any active patient" due to the high degree of variability in how practices defined "active" and how frequently they purged records. Choosing to include only those patients that had been seen in the past twelve months yielded a more optimistic picture relative to compliance than what we might otherwise have obtained.

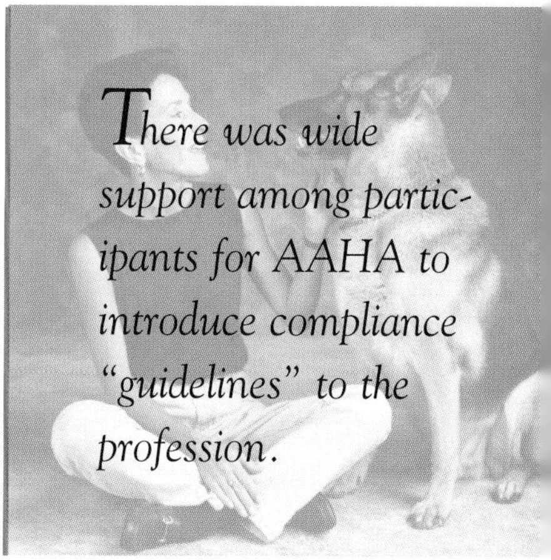

*There was wide support among participants for AAHA to introduce compliance "guidelines" to the profession.*

## Compliance Estimates

The compliance levels estimated by the participants were accurate for some services, poor for others (see Table 6).

TABLE 6

## Estimated and Actual Compliance Rates

| | Compliance Area | Estimated Compliance Rate | Actual Overall Compliance Rate | Estimated: Actual Ratio | |
|---|---|---|---|---|---|
| UNDERESTIMATED COMPLIANCE | Heartworm Testing | 73% | 83% | 88% | |
| | Core Vaccines | 77% | 87% | 88% | |
| | Preanesthetic Testing | 66% | 69% | 97% | |
| OVERESTIMATED COMPLIANCE | Senior Screening | 43% | 34% | 128% | WEIGHTED AVERAGE OVERESTIMATE: 51% |
| | Heartworm Preventive | 70% | 48% | 147% | |
| | Dental Prophylaxis | 54% | 35% | 153% | |
| | Therapeutic Diets | 59% | 21% | 192% | |
| | Overall[1] | 67% | 64% | 105% | |

NOTE 1: Overall average weighted by the number of patients eligible for each service/product.

When asked, "How do you estimate your compliance levels?" 83% indicated they have a "feel," 40% said they track reminder card response rates, and 26% said they estimate via financial and activity tracking. Actual compliance tracking (for at least one area of compliance) was claimed by 22% of respondents.

## Compliance Analysis

Based on data from patients' charts, we analyzed actual compliance rates for six areas: heartworm testing and preventive, dental prophylaxis, therapeutic diets, core vaccines, senior screenings, and preanesthetic testing.

### Heartworm testing and preventive

We used data from the American Heartworm Society and other sources to determine which states were considered to be heartworm endemic, nonendemic, and in transition. Of the 242 practices reporting data, 16% of those reporting from heartworm endemic states self-reported as being in nonendemic heartworm areas; another 8% self-reported as being in states that are in transition. Not surprisingly, the practices that reported an endemic status that was different than the AHS status had significantly lower compliance rates for heartworm testing and preventive.

One hundred percent of the practice teams in endemic states (that also self-reported being in an endemic state) indicated that they routinely recommend heartworm testing and preventive medication. However, the medical-record data revealed compliance rates as shown in Table 7.

TABLE 7

**Compliance Rates for Heartworm Testing and Preventive in Heartworm Endemic States**

| | |
|---|---|
| Percentage of total that had had a heartworm test recommended | 89% |
| Percentage of those that had had a recommendation that were current with testing | 93% |
| Percentage of total that were current with testing | 83% |
| Percentage of total that were current with preventive medication[1] | 48% |
| Percentage of recommended doses purchased[2] | 65% |

NOTE 1: Purchased at least enough preventive in the last twelve months to comply with veterinarian's recommendation.

NOTE 2: Averaged across all dogs in endemic areas.

In some cases, practices were in states generally considered to be endemic, but the team members reported that their practices were in nonendemic or transitional areas. In these practices, only 72% of patients were in compliance with heartworm testing recommendations (versus 93% of the practices whose team members reported that the practice was in an endemic area), and only 50% of total recommended doses were purchased (versus 65% in the practices whose team members reported that the practice was in an endemic area).

## Dental prophylaxis

We asked the study participants to indicate, for each patient record reviewed, the grade of dental disease recorded (Grade 0 through Grade 4). However, *no grade was reported* for 19% of the patients. The lack of a reported dental grade for so many patients may have been an indication that no exam was given or that record keeping was poor. While this uncertainty did not change the recommendation gap and compliance gap calculations for those with known dental disease grades, it did diminish the reliability with which we calculated the percentage of patients that did not receive a needed dental procedure.

As can be seen in Table 8 on page 71, the low level of compliance in dentistry is due in large part to the "recommendation gap." The recommendation gap is largest for patients with Grade 1 dental disease—in 66% of the cases where patients had Grade 1 dental disease, veterinarians did not recommend dental prophylaxis. The American Veterinary Dental College defines quality dental health care as completing a dental prophylactic procedure on any pet with Grade 1–4 dental disease. Obviously, veterinary health care teams have failed to adhere to these recommendations in a great many cases, which has resulted in less-than-the-best care for many patients.

TABLE 8

## Recommendation Gap for Dental Prophylaxis

| | Grade of Dental Disease | | | | | |
|---|---|---|---|---|---|---|
| | None Reported | 0 | 1 | 2 | 3 | 4 |
| Number of Patients | 262 | 440 | 319 | 204 | 103 | 36 |
| Percentage of Graded Patients | N/A | 40% | 29% | 19% | 9% | 3% |
| Percentage of All Patients | 19% | 32% | 23% | 15% | 8% | 3% |
| Percentage That Received Recommendation | 17% | 12% | 34% | 69% | 87% | 94% |
| Percentage That Did Not Receive Recommendation | N/A | N/A | 66% | 31% | 13% | 6% |
| Percentage of Clients That Accepted Recommendation | 56% | 84% | 45% | 43% | 48% | 50% |
| Percentage of Total That Received Dental Procedure | 10% | 10% | 15% | 29% | 42% | 47% |

### Therapeutic diets

As with other services and products we studied, there was a significant "recommendation gap" related to compliance with therapeutic diets. We studied six conditions in dogs: kidney disorder, bladder stones/crystals, food allergy, chronic gastrointestinal disease, acute gastrointestinal disease, and obesity. A therapeutic diet was recommended in 76% of the cases where a diagnosis of one of these six conditions was reported. For the seven conditions we studied in cats (the six listed for dogs plus feline lower urinary tract disease), a recommendation for therapeutic diet was

While the data from this study point to the potential for five-fold improvement in compliance with therapeutic diet recommendations, the real potential for improvement for all diets combined could be as high as twenty-fold.

reported for just 71% of the cases where a diagnosis of one of the conditions was reported.

If a therapeutic diet recommendation was made, clients purchased the recommended food 55% of the time. However, compliance rates declined after the initial diet purchase either because the client did not continue to purchase the diet and/or because the client fed foods in addition to the recommended diet.

As was noted in other modules in this study, client compliance related to therapeutic diets increased when the practice followed up on the recommendations and provided assistance to the clients in order to help the pet make a successful transition to the new diet.

Overall compliance with therapeutic diets in this portion of the study was 21%, which correlates well with the compliance calculated in Module III, which was 19% for dogs and 18% for cats. It is important to note that these compliance figures relate only to the diagnoses we studied.

Based on comparison with industry-supplied data from Module I, we believe that compliance with therapeutic diets for other conditions may be far lower. In fact, when all diagnoses that could benefit from treatment with a therapeutic diet were considered, overall compliance was 5–7%. While the data from this portion of the study alone points to the potential for five-fold improvement, the real potential for improvement for all diets combined could be as high as twenty-fold.

### Core vaccines

Overall compliance for canine and feline core vaccines was quite high at 87%. The compliance rate for puppies and kittens was approximately 5% higher than for adult pets, and the compliance rate for dogs was almost 10% higher than for cats. It should be noted, however, that because all of the pets in this sample were seen within the past twelve months, actual vaccine compliance was very likely overstated.

### Senior screenings

Overall, only 34% of those pets defined by each veterinarian as "senior pets" had had any diagnostic screening procedures performed. These data showed that the biggest cause of low

Veterinarians recommended senior screenings to owners of only 45% of senior dogs and 48% of senior cats.

compliance rates was the lack of a recommendation, not the client's refusal of a recommendation.

Veterinarians recommended senior screenings to owners of only 45% of senior dogs and 48% of senior cats. When senior screenings were recommended, 70% of dog owners and 74% of cat owners accepted the recommendation. Since client acceptance was so high, even a modest increase in the number of times a senior screening was recommended would have resulted in a very large increase in overall compliance. (Specifically, increasing the recommendation rate to 75% would result in an overall compliance rate of 54%.)

### Preanesthetic testing

Protocols in most practices participating in this study required either preanesthetic screening for all anesthesia cases, required it for high-risk patients, or strongly recommended it for all patients. Not surprisingly, therefore, compliance was relatively high at 69%, but it was clearly correlated with the strength of the requirement/recommendation.

About 9% of the practices included in this portion of the study either made preanesthetic screening optional or did not strongly recommend it. The percentage of patients that underwent preanesthetic screening in these practices was less than 10%.

## Other Observations

As noted in other sections of the study, the degree of staff involvement correlated strongly with compliance. Repetition of recommendations (often by a staff member other than the doctor) also had a significant impact on compliance, as did the availability of written materials, either from the practice, AAHA, or the animal health industry.

Monitoring compliance was significantly correlated with increased compliance. In the case of vaccines, for example, most practices employed a somewhat rudimentary means of measuring compliance by tracking responses to reminders. Most practices sent multiple reminders to those clients who were overdue, and many also called their clients if the second or third reminder had not resulted in compliance. It is not surprising, therefore, that vaccine compliance was higher than in any other area we studied.

Overall compliance rates in practices that monitored compliance were higher than in those that did not monitor compliance at all and those that only occasionally monitored compliance. Overall compliance rates for the practices we studied are listed in Table 9.

TABLE 9
**Compliance Rates by Frequency of Compliance Monitoring**

| | Percentage of Cases That Received Recommendation and Accepted Treatment | Percentage of All Cases That Needed Treatment That Received Treatment |
|---|---|---|
| Did Not Monitor Compliance | 67% | 55% |
| Occasionally Monitored Compliance | 70% | 61% |
| Regularly Monitored Compliance | 81% | 78% |

The first column shows the percentage of owners who followed through on a recommendation received. The second column shows the owners who followed through as a percentage of those pets that needed a preventive/treatment/screening (based on age, heartworm endemic status, dental grade, and diagnoses made).

It is interesting to note that the effect of monitoring can be seen at both the "recommendation" level and at the "client acceptance" level but appeared to affect the recommendation gap most. In other words, monitoring compliance had a positive impact on the number of times a recommendation was made.

Practice demographics, including the number of full-time-equivalent veterinarians and gross revenue, did not appear to affect compliance. However, practices with more women veterinarians had slightly (but statistically significantly) higher compliance rates (75%) than practices overall (73%) and than practices in which less than 30% of the veterinarians were women (70%). It was beyond the scope of this study to explore the reasons for those differences.

## Quantifying Opportunities Available by Increasing Compliance Rates

Using key assumptions from a number of other studies (e.g., the total number of cats and dogs and the number of pets receiving veterinary care), we created a model that estimated both the additional number of pets that could receive better care and the additional revenue that could be generated through even modest increases in compliance.

Tables 10 and 11 detail the number of additional pets that could be treated and the additional revenue that could be generated through increases in compliance of ten percentage points and twenty-five percentage points in an average practice (2.2 full-time-equivalent veterinarians and 3,475 active patients). A ten percentage point increase in compliance means that an average practice provides 1,287 more treatments and earns $132,535 in additional revenue. A twenty-five percentage point increase in compliance means that an average practice provides 2,674 more treatments and earns $308,423 in additional revenue.

TABLE 10

## How a 10% Increase in Compliance Affects Revenue[1,2]

| Compliance Area | Current Overall Rate of Compliance | Compliance Goal | Number of Additional Treatments per year | Increase in Revenue Through Improved Compliance |
|---|---|---|---|---|
| Therapeutic Diets | 5% | 15% | 202 | $48,587 |
| Dental Prophylaxis | 29% | 39% | 158 | $31,521 |
| Core Vaccines and Related Exams | 87% | 97% | 337 | $16,294 |
| Senior Screenings | 33% | 43% | 119 | $13,483 |
| Heartworm Preventive | 48% | 58% | 180 | $11,988 |
| Preanesthetic Testing | 63% | 73% | 111 | $6,072 |
| Heartworm Testing | 83% | 93% | 180 | $4,590 |
| Total Improvement | | | 1,287 | $132,535 |

NOTE 1: Calculation includes only the areas we studied, except for diets, which, for the quantification of the opportunity, we expanded beyond the diagnoses studied.

NOTE 2: Revenue opportunity was calculated using average fees compiled from a variety of sources, including *The Veterinary Fee Reference*.

TABLE 11

**How a 25% Increase in Compliance Affects Revenue[1,2]**

| Compliance Area | Current Overall Rate of Compliance | Compliance Goal | Number of Additional Treatments per year | Increase in Revenue Through Improved Compliance |
|---|---|---|---|---|
| Therapeutic Diets | 5% | 30% | 506 | $121,469 |
| Dental Prophylaxis | 29% | 54% | 396 | $79,002 |
| Core Vaccines and Related Exams | 87% | 100% | 441 | $21,322 |
| Senior Screenings | 33% | 58% | 297 | $33,650 |
| Heartworm Preventive | 48% | 73% | 450 | $29,970 |
| Preanesthetic Testing | 63% | 88% | 278 | $15,207 |
| Heartworm Testing | 83% | 100% | 306 | $7,803 |
| Total Improvement | | | 2,674 | $308,423 |

NOTE 1: Calculation includes only the areas we studied, except for diets, which, for the quantification of the opportunity, we expanded beyond the diagnoses studied.

NOTE 2: Revenue opportunity was calculated using average fees compiled from a variety of sources, including *The Veterinary Fee Reference*.

## Module VI: Pet Owner Research

In November 2002, 1,670 surveys were mailed to dog and cat owners by NFO WorldGroup. The NFO panel is balanced relative to United States census data with respect to household income, race, education, age, geographic region, and gender. By the survey deadline four weeks later, 1,003 pet owners had responded, for a response rate of 60%. Participants were asked to focus on a single randomly selected pet (the one whose name was alphabetically first among their pets) when answering pet-related questions.

## Key Findings

We asked several questions to determine whether our study was consistent with previously published pet owner research, and there was a very high correlation. Here is a summary of our findings:

- A total of 74% of all pet owners have taken their pets to see a veterinarian one or more times a year. More than 25% of dogs and 10% of cats were taken to the veterinarian more than once a year.
- Of all respondents, 24% have switched veterinarians during this pet's lifetime. Of those that switched, 75% did so because they moved or sought a more convenient location for service. Consistent with other surveys, 15% of those that switched (4% of all pet owners) did so because the "prior veterinarian was too expensive."
- As with other surveys, pet owners reported a high degree of satisfaction with the service and care received from their veterinarians: 93% of respondents were either "very satisfied" (82%) or "somewhat satisfied" (11%). For the 18% that were less than "very satisfied," cost was the single most frequently cited reason. However, only 8% of clients mentioned cost at all.
- Because work in earlier modules suggested that veterinarians may fail to make health care recommendations due to a perceived lack of willingness (or ability) of the client to pay the cost of that care, clients were asked to indicate which of the following two statements best indicated their preference:

1. I want my veterinarian to tell me about all the recommended treatment options for my pet, even if I may not be able to afford them.

2. I want my veterinarian to tell me only about the recommended treatments for my pet that he or she thinks are not too expensive for me.

Ninety percent of the respondents selected the first statement—they wanted to hear about all of the options, irrespective of the cost. Only 10% selected the second statement.

- Seventy percent of dog owners recalled receiving a recommendation for heartworm preventive. However, the percentages in some endemic states were surprisingly low (e.g., Alabama 43%, South Carolina 50%, Louisiana 71%).
- In response to a question asking where respondents purchased heartworm preventive, 96% indicated that they made the purchase at their veterinarian's office; only 4% purchased by mail or through the Internet.
- Fifty percent of those who received a recommendation for dental care followed through and scheduled the dental procedure for the pet (in close agreement with Module V data). Of the remaining 50%, at least half were likely to schedule the procedure and could have been relatively easily moved to do so if the practice had followed up. Only 14% of those who received a recommendation and had not complied cited cost as the reason (7% of those who received a recommendation).
- Of the pets reported by their owners to have been diagnosed with a specific medical condition, 84% recalled receiving a recommendation for a therapeutic diet. Seventy-two percent purchased the diet, and 60% fed the diet as directed.
- Most pet owners who received a dietary recommendation

remained compliant; if they did not, palatability and acceptability of texture and type (e.g., canned, dry) were the most common reasons. However, 55% of pet owners who fed a therapeutic diet supplemented the recommended diet with other food or treats, corroborating the therapeutic diet compliance rates documented in Modules III and V.

- Only 30% of owners of senior pets indicated that screening lab work had been done. The remaining 70% indicated that they had "never heard of it," "didn't know I should do it," or "my veterinarian never recommended or suggested it." This was a clear indication that the failure of the practice team to make a recommendation was to blame for low compliance.
- Pet owners believed that their veterinarians were highly competent, and veterinarians were their primary source of healthcare information for their pets.
- Pet owners did not think that veterinarians were motivated by money. Almost 75% agreed or completely agreed that their veterinarians made recommendations because they were good for the pet. Less than 10% felt that making more money was the reason veterinarians make health care recommendations.
- In general, pet owners felt that communication with their veterinarians and practice staff were good. While 97% of pet owners agreed or completely agreed that their veterinarian was good at explaining pet health care and health problems to them, almost 60% agreed or completely agreed that their veterinarian or practice did not always make it clear to them how *important* the heath care recommendations were.
- Clients saw the members of the practice staff as knowledgeable and helpful and thought the staff provided good service.
- Over 75% of pet owners wanted their veterinarians or the staff to call them to follow up on the pet's condition after

the pet had had a problem. However, only about 52% agreed or completely agreed with the statement that their veterinary practices had provided that follow-up.

- Pet owners wanted dietary advice, particularly for a sick pet, and were willing to spend more for a recommended diet if it would help their pets. Further, they wanted advice about the best food to feed even when the pet is healthy.
- Pet owners wanted home-care instructions both verbally and in writing, and they saw the practice staff as an important resource for home-care and feeding issues.
- Nearly all pet owners (over 90%) indicated that they wanted to be reminded when they were due for examinations, treatments, vaccinations, and medication refills. An equal number indicated that they did receive such reminders. Many (approximately 72%) also wanted to be called if overdue.
- As with previous pet owner studies, these respondents reported a very high care ethic and were emotionally invested in their pets as family members. They also agreed that veterinary care was affordable, and the majority said they would spend whatever it would take to save a pet's life.

*The majority of pet owners said they would spend whatever it would take to save a pet's life.*

## Cost as a Barrier

Based on the empirical evidence contained in this study, the client is not the major barrier to improving compliance in veterinary medicine, and cost is rarely an issue:

- Only 7% do not comply with dental recommendations due to cost.

- Only 4% abandon recommended diets due to cost.
- Only 5% turn down senior screenings due to cost.

Unfortunately, many veterinarians (and sometimes practice staff) choose to be heavily influenced by the less than 10% of clients for whom cost is ever an issue, rather than focusing on the 90% who want the best health care for their pets. This study has shown that the profession does not consistently and confidently make recommendations for care that is truly in the best interest of patients. The current quality of care is less than pet owners want and less than the profession wants to provide.

*Unfortunately, many veterinarians (and sometimes practice staff) choose to be heavily influenced by the less than 10% of clients for whom cost is ever an issue, rather than focusing on the 90% who want the best health care for their pets.*

## Conclusion

Compliance with veterinary health care standards is low, but we believe that it can be improved. Carefully consider the following statements:

- You and your staff strive to provide a high quality of care.
- You and your staff have considerable influence over the compliance levels you achieve.
- Pet owners want the best care possible for their pets.
- Your efforts to improve compliance will be cost effective.

Commit now to do the very best you can for your patients. Your revenue will increase, your clients will be more satisfied, and your patients will be healthier!